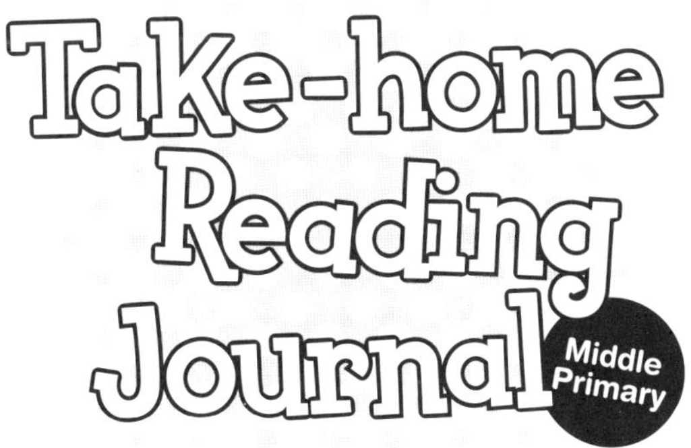

My name is

My class is

These are my friends and me ...

How to use this book

This journal is a record of a child's take-home reading throughout the year. It is designed so that schools may use it in different ways. For example, students may or may not begin each week on a new journal page, and schools may or may not use the reading records as a way of communicating with carers. It is recommended that teachers explain to parents at the beginning of the year how they would like the book to be used, and that parents understand that not all parts of the book must be used.

Journal pages

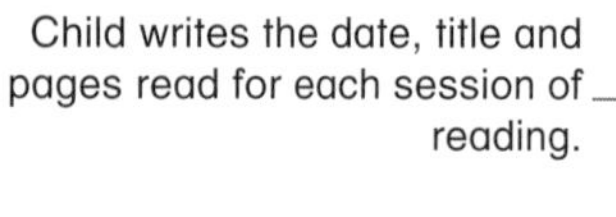

Child writes the date, title and pages read for each session of reading.

Child reflects on the week's reading, and completes a sentence about something they have read.

Child records some new, exciting or unusual words they have come across in their reading.

Teacher comments to carer and child each week (optional).

Carer writes a comment to the child once a week.

Milestones

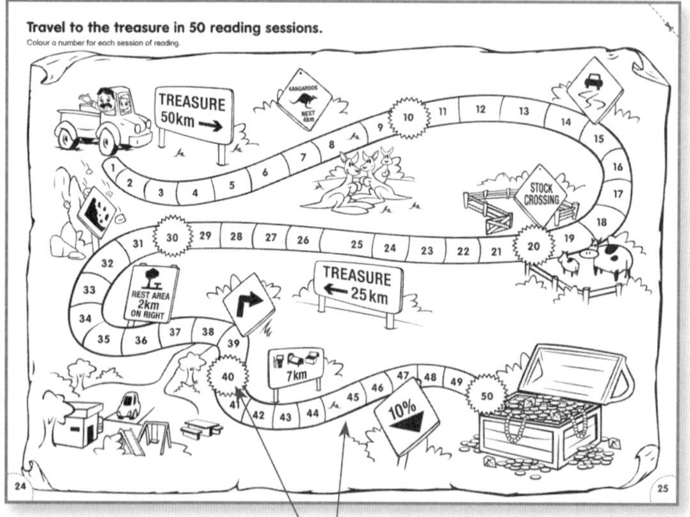

Teachers can decorate points along the way or milestones with congratulatory stickers and stamps.

Child colours a number for each session of take-home reading to reach the treasure (pages 24–25). On pages 36–37, 48–49 and 60–61 they continue to colour in numbers to reach milestones of 100, 150 and 200 sessions of reading.

Premier's Reading Challenge

On pages 70–71, the child can fill in books read for the Premier's Reading Challenge.

At the end of the year, the child receives a certificate stating how many sessions of take-home reading they have completed in the year (see page 72).

Getting started with take-home reading

Reading at home is a time for parents (or caregivers) and children to read together. Children practise the reading skills they are learning at school when they read to their carers. Children will bring home a range of fiction and non-fiction books to read.

Reading at home is an important and enjoyable way to help your child to learn. Children who read regularly at home learn to read more quickly and easily.

- Set aside 15 to 20 minutes each night.
- Find a comfortable, quiet place to read.
- Sit side by side so that you can both easily see the book.
- Listen to your child read. By this stage he or she may be reading part of a longer book to you.
- Talk about what you have read. (See 'What do I ask?' on page 7.)

What is the difference between an early and an independent reader?

Early readers:

- read slowly and deliberately, often focusing on each word, especially when reading new or unfamiliar texts
- recognise many words and use various strategies (e.g. meaning and visual clues, and word structure) to work out unknown words
- sometimes comment on and question what they read
- can adapt their strategies to different types of text.

Independent readers:

- read a range of texts fluently and with ease
- use various strategies to approach unfamiliar or specialised texts
- can predict and self-correct if necessary to maintain meaning
- often make connections between what they know and understand, and what is new.

My child is not yet an independent reader. How can I help?

Continue with take-home reading. Children may have a little difficulty, especially with more challenging or less familiar texts. You might help by:

- leafing through the book first and talking about this type of book and the pictures.
- being patient and understanding. Learning to read is a very complex task. It takes time.
- giving your child time to work things out by themselves. (Reading is often about solving problems. Some carers find that it helps if they silently count to 10 before offering any help.)

Talk to your child's teacher about other specific things you could do to help.

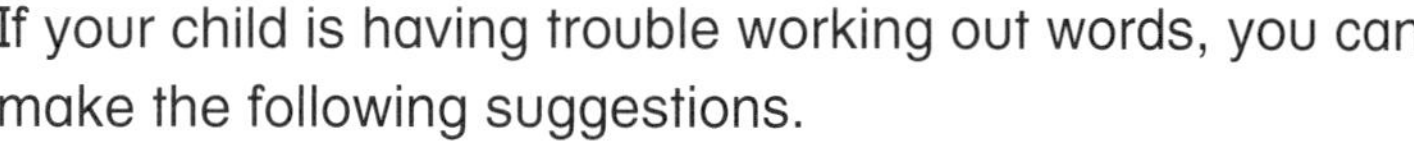

What can I do when my child has trouble with individual words?

If your child is having trouble working out words, you can make the following suggestions.

- *Look at any pictures, diagrams or photos.* These can help with the overall meaning of a page or section, as well as with unfamiliar words.
- *What do you think the word might be? What would sound right there?*
- *Look at the first letter. Can you say part of the word?*
- *Do you know another word that looks like it?* For example, they may know 'throw' looks like 'three'.
- *Keep reading and we can go back to it.* Reading on for the meaning to the end of the phrase or sentence can often help with difficult words.
- *Does it make sense? Does it look right?* Encourage self-correction.

My child is reading independently. Should I still listen to him or her read?

Yes. You will both benefit from continuing with the take-home reading. Your child's literacy skills in oral language will continue to improve through reading aloud. His or her comprehension skills and knowledge of different text types will improve through appropriate questioning about the reading. You will continue to stay in touch with your child's literacy development by hearing him or her read. Of course, you may choose to listen for a short while and then leave him or her to read silently; and you will want to encourage more silent reading.

Understanding the features of different text types is an important aspect of literacy. Discuss these with your child.

Text type	Features	Examples
Narrative	tells a story or gives an account of real or imaginary events	*Amy's Adventures in Amazonia* *Cinderella*
Procedure	describes how to do something	*Rules for Playing Checkers* *Granny Annie's Tomato Soup*
Report	a factual text that describes the way things are in the natural or social world	*Australian Bases in Antarctica*
Recount	retells the details of an event or experience or a person's life	*My Mountain-climbing Adventure* *Captain Cook's Childhood*
Explanation	describes something and tells how it works or why it happens	*How a CD Player Works* *What are Hailstones?*
Exposition	explores arguments about an issue	*Should Children Have Mobile Phones?* *Soccer vs Aussie Rules*

Use questions such as these to discuss the story, article or other piece of text.

Questions for fiction or non-fiction

- *What sort of book or article is it? What text type?*
- *Who is the author? Have you read anything else by this author?*
- *Is it illustrated? Who by? Have you read or seen any other books illustrated by this person?*
- *Where are the author and illustrator from?*
- *Why do you think the author wrote the story or article?*
- *What surprising or interesting things did you find out by reading this text?*
- *Have you read anything like this before?*
- *Tell me three words that describe how you felt after reading this.*

Questions for fiction

- *What sort of writing is it (for example, a mystery, an adventure, a funny story, a romance)?*
- *Who are the main characters?*
- *If this story was made into a movie, who do you think would play ____ ?*

Questions for non-fiction

- *How do different features of the book or article (for example, headings, contents, diagrams, glossary, captions, labels, numbered instructions) help you to understand it?*
- *What did you learn from reading this text?*

At this stage, an encouraging weekly comment is appropriate. Your comments could be directed to your child or to the teacher. Here are some ideas:

- *I am proud of you for reading such a big book.*
- *Good on you for remembering to read on when you get stuck.*
- *You are a great recipe reader, and a great cookie baker!*
- *Jess and I have really enjoyed talking about what might happen next.*
- *Dylan read FIVE chapters this week! Brilliant!*
- *We learnt a lot about dolphins and whales this week.*

Other things to do at home to help your child's reading and writing

- Continue to read to your child every day. Read a variety of books, from short stories and information books to longer books that you can read as a serial.
- Visit your local library regularly, and encourage your child to borrow both familiar and new books, authors, and types of texts.
- Read and discuss reading materials such as newspapers, recipes, TV guides, calendars, magazines and puzzle books.
- Look for simple recipes in library books or magazines, and use them to cook with your child.
- Encourage your child to read to younger brothers and sisters, other family members and friends.
- Let your child see you reading and enjoying it.
- Have your child help you with shopping lists, phone messages, letters and greeting cards.
- Encourage your child to read and follow instructions for playing games, building toys or putting new household goods together.
- Read books that have been made into films, and discuss the differences between the two versions.
- Play word games together, at home or in the car.
- Listen to recorded stories, poems and songs on CD, at home or in the car.
- Buy books and word games as presents.
- Talk about the plot and the ending of stories; discuss alternatives or 'what ifs'.
- Make up new verses of favourite poems and songs, or new versions of fairy stories.

Great books to read

Here are some ideas for books that are suitable for different reading abilities in the middle years of primary school. If you are searching for other books for your child to read at home, look for these at your local library or bookshop.

First chapter books

Aussie Nibbles (various authors, Penguin/Puffin)
Boyz Rule series (written by Felice Arena and Phil Kettle, Macmillan Australia)
Too Cool series (written by Phil Kettle and illustrated by Craig Smith, Scholastic Australia)
Solos (various authors, Scholastic Australia)
Squeak Street books (written by Emily Rodda and illustrated by Andrew McLean, Working Title Press)

Longer chapter books

Aussie Angels series (written by Margaret Clark, Hodder Headline Australia)
Aussie Bites (various authors, Penguin/Puffin)
Finn Family Moomintroll (written and illustrated by Tove Jansson, Puffin, 1961)
Go Girl series (written by various authors and illustrated by Ash Oswald, Hardie Grant Egmont)
Making Tracks series (various authors and illustrators, National Museum of Australia Press)
Tashi series (written by Anna and Barbara Fienberg and illustrated by Kim Gamble, Allen & Unwin)
Worst Witch series (written by Jill Murphy, Puffin)

Advanced chapter books

The BFG (written by Roald Dahl and illustrated by Quentin Blake, Puffin, 1982)
Finders Keepers (written by Emily Rodda, Omnibus Books, 1992)
45 + 47 Stella St: and Everything That Happened (written and illustrated by Elizabeth Honey, Allen & Unwin, 2000)
Rain May and Captain Daniel (written by Catherine Bateson, University of Queensland Press, 2002)

Ratbags and Rascals (written by Robin Klein and illustrated by Alison Lester, Dent, 1984)
Specky McGee series (written by Felice Arena and Garry Lyon, Puffin)

Picture books for middle primary readers

The Eleventh Hour (written and illustrated by Graeme Base, Viking Puffin, 1988)
Eyes in the Night (written by Jan Ramage and illustrated by Laura Peterson, University of Western Australia Press, 2004)
The Lost Thing (written and illustrated by Shaun Tan, Lothian, 2000)
The Rainbow (written by Gary Crew and Gregory Rogers, Lothian, 2001)
Squids Will Be Squids (written and illustrated by Jon Scieszka, Viking Penguin, 1998)
Storm Boy (written by Colin Thiele, Rigby, first published 1963, various editions with various illustrations or photographs)

Non-fiction books

Animal Architects (written and illustrated by John Nicholson, Allen & Unwin, 2003)
Eyewitness Guides series (various authors, Dorling Kindersley)
Gogo Fish! The Story of the Western Australian State Fossil Emblem (written by John Long, illustrated by Jill Ruse and John Long, Western Australian Museum, 2004)
My Place (written by Nadia Wheatley and illustrated by Donna Rawlins, Collins Dove, 1987)
Queenie: One Elephant's Story (written by Corinne Fenton and illustrated by Peter Gouldthorpe, Black Dog Books, 2006)
Tough Stuff (written by Kirsty Murray, Allen & Unwin, 1999)

150 instant words

This list contains 150 high frequency words. You will already know many of them. It is important to learn to recognise them all instantly.

a	give	my	them
about	go	name	then
after	good	new	there
all	great	no	these
also	had	not	they
am	has	now	thing
an	have	number	think
and	he	of	this
any	help	old	three
are	her	on	through
around	him	one	time
as	his	only	to
at	how	or	too
back	I	other	two
be	if	our	up
been	in	out	use
before	into	over	very
boy	is	part	want
but	it	people	was
by	its	place	water
call	just	right	way
came	know	said	we
can	like	same	were
come	line	say	what
could	little	sea	when
day	live	sentence	where
did	long	she	which
do	look	show	who
down	made	small	will
each	make	so	with
find	man	some	word
first	many	sound	work
follow	may	take	would
for	me	tell	write
form	mean	than	year
from	more	that	you
get	most	the	your
	much	their	

150 useful words

These words are grouped into themes. Look at them regularly and become familiar with as many as possible.

Time

- second
- minute
- hour
- day
- week
- month
- year
- century

Days

- Monday
- Tuesday
- Wednesday
- Thursday
- Friday
- Saturday
- Sunday

Months

- January
- February
- March
- April
- May
- June
- July
- August
- September
- October
- November
- December

Seasons

- spring
- summer
- autumn
- winter

Family

- aunt
- baby
- boy
- brother
- child
- children
- cousin
- daughter
- father
- girl
- grandfather
- grandmother
- mother
- parent
- sister
- son
- uncle

Body

- ankle
- arm
- back
- bones
- bottom
- brain
- cheek
- chest
- chin
- elbow
- eye
- face
- finger
- foot
- hand
- head
- heart
- heel
- knee
- leg
- lip
- mouth
- neck
- nose
- shoulder
- skin
- stomach
- teeth
- thumb
- toe
- tongue
- waist
- wrist

Places

- airport
- beach
- camping
- city
- country
- creek
- farm
- hills
- library
- movies
- museum
- oval
- park
- picnic
- pool
- shops
- space
- zoo

Transport

- bicycle
- boat
- bus
- car
- motorbike
- plane
- rocket
- scooter
- ship
- train
- tram
- truck
- van
- vehicle

Colour

- black
- blue
- bright
- brown
- dark
- green
- grey
- light
- orange
- pink
- purple
- red
- violet
- white
- yellow

Technology

- computer
- CD-ROM
- disk
- Internet
- keyboard
- mobile
- phone
- radio
- recorder
- screen
- video
- website

Date	Title	Pages

Complete this sentence about one item you read. (Tick it on your list.)

The most exciting part was ______________________________

Write new words you read this week.

Word	Meaning

Carer's comment to child	Teacher's comment to carer and child

How do you spell "hungry horse" in four letters?

M T G G.

Date	Title	Pages

Complete this sentence about one item you read. (Tick it on your list.)

I was really surprised about

Write new words you read this week.

Word	Meaning

Carer's comment to child	Teacher's comment to carer and child

Date	Title	Pages

Complete this sentence about one item you read. (Tick it on your list.)

I learned this interesting fact: ____________________

Write new words you read this week.

Word	Meaning

Carer's comment to child	Teacher's comment to carer and child

Date	Title	Pages

Complete this sentence about one item you read. (Tick it on your list.)

My favourite character was ______________________ because ______________________

Write new words you read this week.

Word	Meaning

Carer's comment to child	Teacher's comment to carer and child

Date	Title	Pages

Complete this sentence about one item you read. (Tick it on your list.)

I especially enjoyed reading this because ____________________

____

Write new words you read this week.

Word	Meaning

Carer's comment to child	Teacher's comment to carer and child

What starts with P, ends in E, and has a lot of letters?

A post office.

Date	Title	Pages

Complete this sentence about one item you read. (Tick it on your list.)

I would like to know more about

Write new words you read this week.

Word	Meaning

Carer's comment to child	Teacher's comment to carer and child

Date	Title	Pages

Complete this sentence about one item you read. (Tick it on your list.)

I would like to tell the author ______________________________

____

Write new words you read this week.

Word	Meaning

Carer's comment to child	Teacher's comment to carer and child

What did the zero say to the eight?

Nice belt!

Date	Title	Pages

Complete this sentence about one item you read. (Tick it on your list.)

I would tell my friends to read this because ______

Write new words you read this week.

Word	Meaning

Carer's comment to child	Teacher's comment to carer and child

Date	Title	Pages

Complete this sentence about one item you read. (Tick it on your list.)

My favourite illustration, diagram or photo was

Write new words you read this week.

Word	Meaning

Carer's comment to child	Teacher's comment to carer and child

Date	Title	Pages

Complete this sentence about one item you read. (Tick it on your list.)

I think this would make a great movie because

Write new words you read this week.

Word	Meaning

Carer's comment to child	Teacher's comment to carer and child

Travel to the treasure in 50 reading sessions.

Colour a number for each session of reading.

9
10
11
12
13
14
15
16
17
18
19
20
21
22
23
24
25
STOCK
CROSSING
ASURE
25 km
5
46
47
48
49
50
10%

Date	Title	Pages

Complete this sentence about one item you read. (Tick it on your list.)

I was really surprised about ____________________

____

Write new words you read this week.

Word	Meaning

Carer's comment to child	Teacher's comment to carer and child

Why was 6 sad?

Because 7 8 9.

Date	Title	Pages

Complete this sentence about one item you read. (Tick it on your list.)

I would like to know more about

Write new words you read this week.

Word	Meaning

Carer's comment to child	Teacher's comment to carer and child

Date	Title	Pages

Complete this sentence about one item you read. (Tick it on your list.)

The most exciting part was ______________________________

____

Write new words you read this week.

Word	Meaning

Carer's comment to child	Teacher's comment to carer and child

Date	Title	Pages

Complete this sentence about one item you read. (Tick it on your list.)

I would like to tell the author ______________________________

Write new words you read this week.

Word	Meaning

Carer's comment to child	Teacher's comment to carer and child

Date	Title	Pages

Complete this sentence about one item you read. (Tick it on your list.)

I especially enjoyed reading this because ______________________________

____

Write new words you read this week.

Word	Meaning

Carer's comment to child	Teacher's comment to carer and child

When were there only three vowels in the alphabet?

Before U and I were born.

Date	Title	Pages

Complete this sentence about one item you read. (Tick it on your list.)

I would tell my friends to read this because

Write new words you read this week.

Word	Meaning

Carer's comment to child	Teacher's comment to carer and child

Date	Title	Pages

Complete this sentence about one item you read. (Tick it on your list.)

My favourite character was ____________________ **because**

____

Write new words you read this week.

Word	Meaning

Carer's comment to child	Teacher's comment to carer and child

2 1 3

Date	Title	Pages

Complete this sentence about one item you read. (Tick it on your list.)

I think this would make a great movie because ______

Write new words you read this week.

Word	Meaning

Carer's comment to child	Teacher's comment to carer and child

Date	Title	Pages

Complete this sentence about one item you read. (Tick it on your list.)

My favourite illustration, diagram or photo was ______________________

____

Write new words you read this week.

Word	Meaning

Carer's comment to child	Teacher's comment to carer and child

How long will the next bus be?

About five metres.

Date	Title	Pages

Complete this sentence about one item you read. (Tick it on your list.)

I learned this interesting fact:

Write new words you read this week.

Word	Meaning

Carer's comment to child	Teacher's comment to carer and child

Swim to the finish line in 50 reading sessions.

Colour a number for each session of reading.

55
56
57
58
59
60
66
65
64
63
62
61
76
77
78
79
80
85
84
83
82
81
97
98
99
100
1
2
3

Date	Title	Pages

Complete this sentence about one item you read. (Tick it on your list.)

I would like to know more about ____________________

Write new words you read this week.

Word	Meaning

Carer's comment to child	Teacher's comment to carer and child

Date	Title	Pages

Complete this sentence about one item you read. (Tick it on your list.)

I learned this interesting fact:

Write new words you read this week.

Word	Meaning

Carer's comment to child	Teacher's comment to carer and child

Date	Title	Pages

Complete this sentence about one item you read. (Tick it on your list.)

I think this would make a great movie because

Write new words you read this week.

Word	Meaning

Carer's comment to child	Teacher's comment to carer and child

What three letters does a robber hate to hear?

I C U.

Date	Title	Pages

Complete this sentence about one item you read. (Tick it on your list.)

I would like to tell the author

Write new words you read this week.

Word	Meaning

Carer's comment to child	Teacher's comment to carer and child

Date	Title	Pages

Complete this sentence about one item you read. (Tick it on your list.)

I was really surprised about ____________________

Write new words you read this week.

Word	Meaning

Carer's comment to child	Teacher's comment to carer and child

Date	Title	Pages

Complete this sentence about one item you read. (Tick it on your list.)

I would tell my friends to read this because

Write new words you read this week.

Word	Meaning

Carer's comment to child	Teacher's comment to carer and child

Date	Title	Pages

Complete this sentence about one item you read. (Tick it on your list.)

I especially enjoyed reading this because

Write new words you read this week.

Word	Meaning

Carer's comment to child	Teacher's comment to carer and child

Where does Thursday come before Wednesday?

In the dictionary.

Date	Title	Pages

Complete this sentence about one item you read. (Tick it on your list.)

The most exciting part was

Write new words you read this week.

Word	Meaning

Carer's comment to child	Teacher's comment to carer and child

Date	Title	Pages

Complete this sentence about one item you read. (Tick it on your list.)

My favourite illustration, diagram or photo was ______________________

____

Write new words you read this week.

Word	Meaning

Carer's comment to child	Teacher's comment to carer and child

Date	Title	Pages

Complete this sentence about one item you read. (Tick it on your list.)

My favourite character was ______________________ because ______________________

Write new words you read this week.

Word	Meaning

Carer's comment to child	Teacher's comment to carer and child

Climb to the rooftop in 50 reading sessions.

Colour a number for each session of reading.

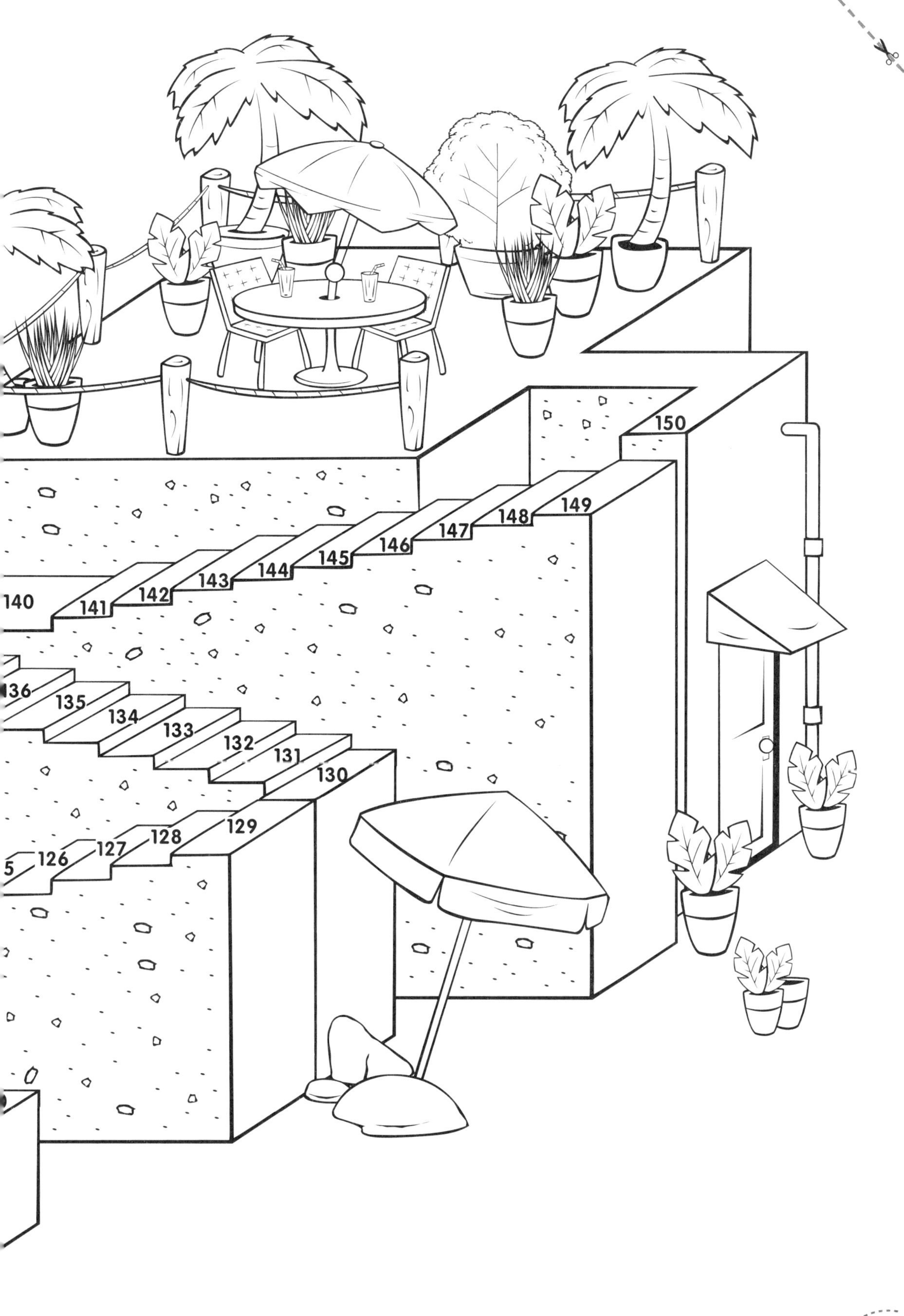
126
127
128
129
130
131
132
133
134
135
136
140
141
142
143
144
145
146
147
148
149
150

Date	Title	Pages

Complete this sentence about one item you read. (Tick it on your list.)

I would like to tell the author ______________________________

____

Write new words you read this week.

Word	Meaning

Carer's comment to child	Teacher's comment to carer and child

What is black and white and red all over?

A sunburnt zebra.

Date	Title	Pages

Complete this sentence about one item you read. (Tick it on your list.)

The most exciting part was

Write new words you read this week.

Word	Meaning

Carer's comment to child	Teacher's comment to carer and child

Date	Title	Pages

Complete this sentence about one item you read. (Tick it on your list.)

I would like to know more about ______________________________

____

Write new words you read this week.

Word	Meaning

Carer's comment to child	Teacher's comment to carer and child

Date	Title	Pages

Complete this sentence about one item you read. (Tick it on your list.)

My favourite illustration, diagram or photo was

Write new words you read this week.

Word	Meaning

Carer's comment to child	Teacher's comment to carer and child

Date	Title	Pages

Complete this sentence about one item you read. (Tick it on your list.)

I learned this interesting fact: ______________________________

____

Write new words you read this week.

Word	Meaning

Carer's comment to child	Teacher's comment to carer and child

Yes, but what is black and white and read all over?

A newspaper.

Date	Title	Pages

Complete this sentence about one item you read. (Tick it on your list.)

I think this would make a great movie because

Write new words you read this week.

Word	Meaning

Carer's comment to child	Teacher's comment to carer and child

Date	Title	Pages

Complete this sentence about one item you read. (Tick it on your list.)

I was really surprised about

Write new words you read this week.

Word	Meaning

Carer's comment to child	Teacher's comment to carer and child

Date	Title	Pages

Complete this sentence about one item you read. (Tick it on your list.)

I especially enjoyed reading this because

Write new words you read this week.

Word	Meaning

Carer's comment to child	Teacher's comment to carer and child

Date	Title	Pages

Complete this sentence about one item you read. (Tick it on your list.)

I would tell my friends to read this because ______________________

Write new words you read this week.

Word	Meaning

Carer's comment to child	Teacher's comment to carer and child

Date	Title	Pages

Complete this sentence about one item you read. (Tick it on your list.)

My favourite character was ______________________ because ______________________

Write new words you read this week.

Word	Meaning

Carer's comment to child	Teacher's comment to carer and child

Fly around the Pacific Ocean in 50 reading sessions.

(The short-tailed shearwater does this trip every year!)

Colour a number for each session of reading.

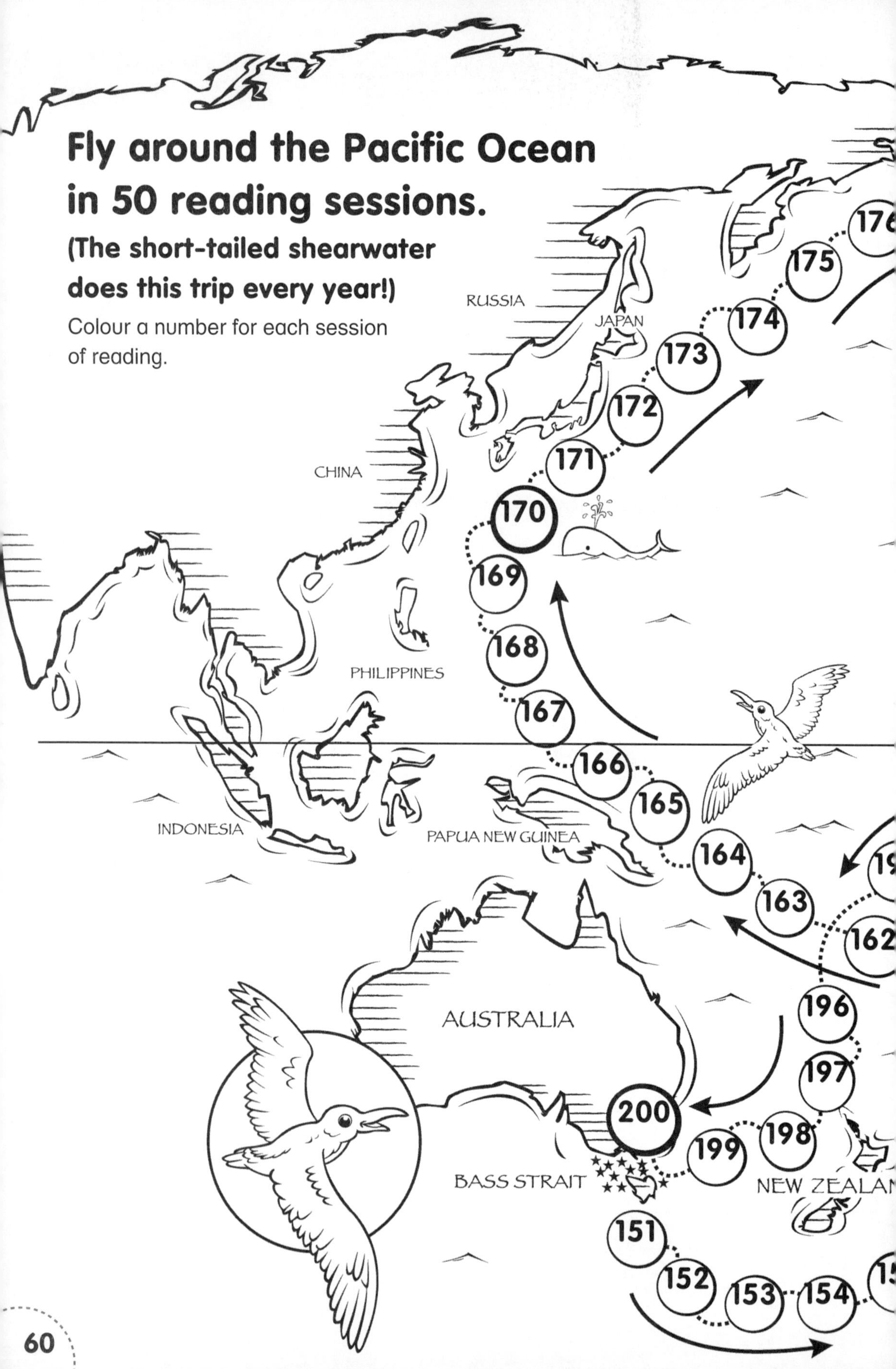

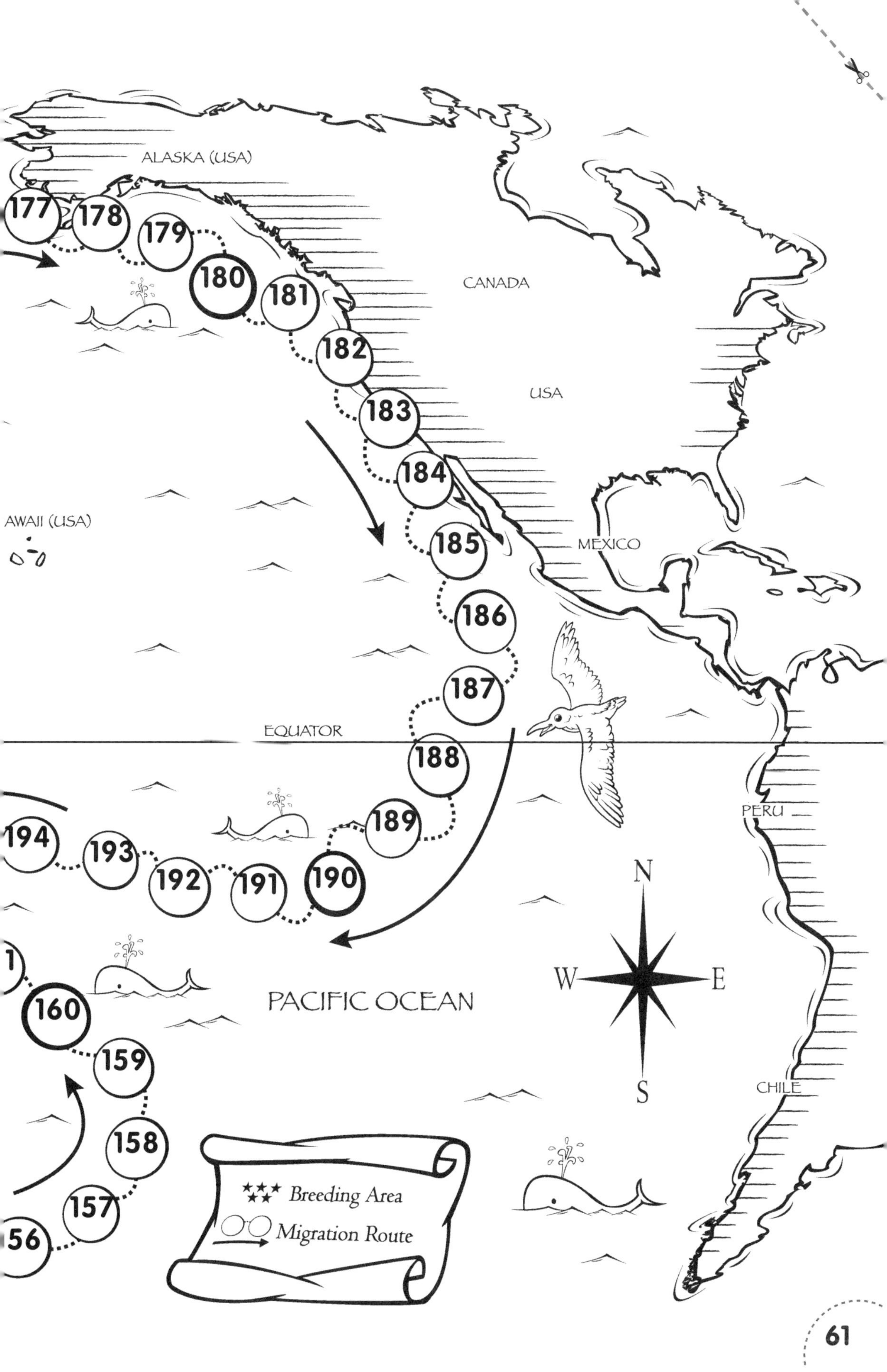
ALASKA (USA)
CANADA
USA
MEXICO
AWAII (USA)
EQUATOR
PERU
CHILE
PACIFIC OCEAN
N
S
W
E
177
178
179
180
181
182
183
184
185
186
187
188
189
190
191
192
193
194
160
159
158
157
56
Breeding Area
Migration Route

معلومات خاصّة بالأهل

- شجّع الطفل على المباشرة بالقراءة.
- أعط الطفل الوقت الكافي ليرتّب أمره بنفسه.
- اسأله على سبيل المثال «ما هو برأيك معنى هذه الكلمة؟»
- اسأله «هل تعرف كلمة أخرى مشابهة؟»
- قل له «حسناً تابع القراءة وسوف نعود إلى هذه الكلمة»
- اسأله «هل لهذه الكلمة معنى؟ هل تبدو لك كلمة صحيحة؟»
- ساعد طفلك مرّة تلو الأخرى على التعرّف إلى الكلمات الرائجة المشار إليها في الصفحات 12 و13 في هذا الكتاب.
- عندما يصبح الطفل أكثر استقلاليّة، قد تجد من المناسب مثلاً أن تستمع إلى بعض قراءاته وتتركه بعدها ليكمل القراءة بنفسه بصوت خافت.

أمور أخرى ممكن اعتمادها من أجل مساعدة الطفل على القراءة والكتابة

- إقرأ كلّ يوم لطفلك. واعتمد قراءة كتب متنوّعة.
- ناقش الكتاب خلال قراءته.
- قم بزيارة المكتبة وشجّع طفلك على استعارة كتب مألوفة أو جديدة وأنواع نصوص لمجموعة معيّنة من الكتّاب.
- إقرأ وناقش مواد القراءة مثل الصحف والوصفات ودليل البرامج التلفزيونيّة والروزنامات والمجلاّت وكتب الأحاجيpuzzle book
- شجّع طفلك على القراءة لأخوته وأخواته الصغار وأصدقائه.
- إجعل طفلك يراك تقرأ وتتمتّع بقراءاتك.
- إجعل طفلك يساعدك في كتابة الأغراض المنزلية التي يجب شراؤها، وبطاقات المعايدة والبطاقات البريديّة والرسائل الهاتفيّة.
- شجّع طفلك على قراءة كتيّب الألعاب أو كتيّب تركيبها أو ترتيب الأغراض المنزليّة واتباع تعليماته.
- شارك طفلك بألعاب الكلمات، واستمع معه في المنزل أو السيارة إلى قصص أو قصائد أو أغان مسجّلة.
- ناقش حبكة القصّة ونهايتها والخيارات الأخرى التي قد تغيّر مجرى القصّة مثلاً «ماذا لو حصل هذا أو ذاك الأمر».
- اخترع مقاطع جديدة من القصائد أو الأغاني المفضّلة أو روايات أخرى لقصص خرافيّة.

كتب مميّزة للقراءة

إذا كنت تبحث عن كتب جديدة للقراءة في المنزل، تجد في الصفحتين 10 و11 من هذا الكتاب لائحة مناسبة.

- كتب الفصل الأوّل
- كتب الفصل الأطول
- كتب الفصل المتقدّم
- كتب مع صور للقرّاء المبتدئين والمتوسّطين
- كتب غير روائيّة

معلومات خاصّة بالأهل

كتاب للقراءة: من المدرسة إلى المنزل

يشكّل دفتر اليوميات هذا سجلاً تدوّن فيه قراءات طفلك للكتب التي يجلبها من المدرسة إلى المنزل للقراءة خلال العام الدراسي. يشجّع جلب كتاب للقراءة من المدرسة إلى المنزل الأهل والطفل على حدّ سواء على تكريس بعض الوقت للقراءة معاً. بهذه الطريقة ومن خلال القراءة بصوت عال أمام أبويه، يتدرّب الطفل على أساليب القراءة التي تعلّمها في المدرسة. الأطفال الذين يقرؤون بشكل منتظم في المنزل، يتعلّمون القراءة أسرع وأسهل.

كيفيّة ملء دفتر اليوميّات

- العودة إلى الصفحة 14. صفحات مماثلة مخصّصة لكلّ أسبوع.
- في الفقرة أعلاه (Date, Title , Pages)، يكتب طفلك التاريخ والعنوان وعدد الصفحات التي قرأها خلال كلّ حصّة قراءة.
- يسترجع طفلك القراءات التي قام بها خلال الأسبوع و يكتب جملة كاملة عن شيء ممّا قرأ. (Complete this sentence...)
- في الفقرة الثالثة، يسجّل طفلك كلمات جديدة مثيرة أو غير عاديّة صادفته خلال قراءاته. (Write new words...)
- في الفقرة في الأسفل جهة اليسار، أكتب تعليقك على سلوك طفلك مرّة في الأسبوع. أمثلة عن التعاليق الممكنة:
  - I am proud of you for reading such a big book.
    أنا فخور بك لقراءتك هذا الكتاب الكبير
  - Good on you for reading on when you get stuck!
    حسناً فعلت بمتابعة القراءة بالرغم من الصعوبة التي واجهتك.
  - Well done! حسناً فعلت بنيّ!
  - We learnt a lot about dolphins and whales this week.
    تعلّمنا الكثير عن الدلافين والحيتان هذا الأسبوع.
  - You are a great recipe reader, and a great cookie baker!
    أنت بارع في قراءة الوصفات وبارع أكثر في تحضير الكعك!
- في الأسفل الجهة اليمنى، يقوم المعلّم/المعلّمة بتدوين ملاحظاته على سلوك طفلك خلال الأسبوع.
- في الصفحات 5–24،7–36،9–48،1–60، يلوّن طفلك أو يدوّن رقماً لكلّ ليلة قراءة حتّى يبلغ عدد ليلات القراءة 50، 100، 150 و200 على مدى العام كلّه.

(ملاحظة: قد لا تعتمد المدارس دوماً دفاتر اليوميات هذه بالطريقة نفسها، كما قد تستثني بعض المدارس أقساماً معيّنة منها.)

كيف نبدأ؟

- تخصيص 15 إلى 20 دقيقة من الوقت كلّ ليلة.
- توفير مكان مريح وساكن من أجل القراءة.
- الجلوس جنباً إلى جنب حتّى تستطيعان معاً متابعة القراءة.
- إستمع إلى طفلك وهو يقرأ. في هذه المرحلة، قد يتوصّل طفلك لقراءة جزء أكبر من الكتاب.
- مناقشة ما تمّت قراءته.

كيف تساعد طفلك في القراءة؟

- كن صبوراً ومتفهّماً. فتعلّم القراءة مهمّة معقّدة جدّاً وتستهلك وقتاً.
- قلّب صفحات الكتاب وناقش نوعه والصور والرسوم الموجودة فيه.

Informasi untuk orang tua dalam bahasa Indonesia

Bahan bacaan untuk dibawa pulang

Jurnal ini merupakan catatan bahan bacaan anak Anda untuk dibawa pulang, selama satu tahun. Ia akan membawa pulang buku ke rumah untuk dibaca pada kebanyakan waktu sore tahun sekolah. Bahan bacaan yang dibawa pulang digunakan bagi memberikan dorongan kepada orang tua dan anak-anak untuk meluangkan waktu bagi membaca bersama. Anak-anak berlatih menggunakan keterampilan membaca yang sedang dipelajari di sekolah sewaktu membacakan kepada orang tuanya. Anak-anak yang sering membaca di rumah belajar membaca dengan lebih cepat dan lebih mudah.

Cara untuk mengisi jurnal

- Lihat halaman 14. Ada halaman seperti ini untuk setiap minggu.
- Di bagian atas (Date, Title, Pages), anak Anda mencatatkan tanggal, judul dan halaman yang dibaca untuk setiap waktu membaca.
- Sekali seminggu, anak Anda meninjau apa yang telah dibaca minggu itu, dan melengkapi kalimat tentang sesuatu yang telah dibaca. (Complete this sentence …)
- Di bagian ketiga, anak Anda mencatatkan beberapa kata yang baru, menarik atau kurang umum, yang ditemui dalam bahan bacaannya. (Write new words …)
- Di bagian kiri bawah, Anda membuat komentar bagi anak Anda sekali seminggu. Contoh komentar adalah:
  - I am proud of you for reading such a big book.
    Saya bangga melihat kamu membaca buku yang begitu tebal.
  - Good on you for reading on when you get stuck!
    Selamat karena tetap membaca walaupun menghadapi kesulitan!
  - Well done! Bagus sekali!
  - We learnt a lot about dolphins and whales this week.
    Kita mempelajari banyak hal tentang lumba-lumba dan ikan paus minggu ini
  - You are a great recipe reader, and a great cookie baker!
    Kamu pandai membaca resep, dan pandai memasak kue!
- Di sebelah kanan bawah, guru memberikan komentar kepada Anda dan anak Anda setiap minggu.
- Pada halaman 24–5, 36–7, 48–9 and 60-1,anak Anda mewarnai atau mengisi satu nomor untuk setiap malam ia membaca untuk mencapai 50, 100, 150 dan 200 malam membaca dalam satu tahun.

(Perhatian: setiap sekolah mungkin tidak menggunakan jurnal ini dengan cara yang sama dan di beberapa sekolah, bagian tertentu dari jurnal mungkin tidak digunakan.)

Cara untuk mulai

- Luangkan waktu 15 sampai 20 menit setiap malam.
- Pergilah ke tempat yang nyaman dan senyap untuk membaca.
- Duduk bersebelahan supaya Anda berdua dapat melihat buku dengan mudah.
- Dengarkan anak Anda membaca. Pada tahap ini ia harus membacakan bagian dari buku yang lebih panjang kepada Anda.
- Bicarakan apa saja yang telah dibaca.

Cara untuk membantu anak Anda membaca

- Bersabarlah dan memahami. Belajar membaca merupakan hal yang rumit sekali dan memakan waktu.
- Lihatlah buku dengan cepat dan bicaralah tentang jenis buku dan gambar, bagan atau foto.

Informasi untuk orang tua dalam bahasa Indonesia

- Berikan dorongan kepada anak Anda untuk 'coba'.
- Berikan waktu kepada anak Anda untuk berpikir sendiri.
- Katakan 'Kamu rasa apa arti kata itu? Apa kata lain yang dapat dipakai di sini?'
- Katakan 'Apakah ada kata lain yang tampaknya serupa?'
- Katakan 'Tetap membaca dan kita bisa kembali ke kata ini nanti.'
- Katakan 'Apakah masuk akal? Apakah tampaknya benar?'
- Lama-kelamaan, bantulah anak Anda mengenal dengan segera kata-kata yang sering dipakai, yang dinyatakan pada halaman 12 dan 13 dari buku ini.
- Sewaktu anak Anda makin mandiri, Anda mungkin memilih untuk mendengarkannya membaca selama waktu yang singkat dan kemudian membiarkannya membaca sendiri.

Hal-hal lain yang dapat dilakukan di rumah untuk membantu anak Anda membaca dan menulis

- Bacakan kepada anak Anda setiap hari. Bacalah berbagai jenis buku.
- Bicaralah tentang buku tersebut sambil membaca.
- Kunjungi perpustakaan setempat, dan berikan dorongan kepada anak Anda untuk meminjam buku yang biasa maupun buku baru dan jenis teks oleh berbagai pengarang.
- Bacalah dan diskusikan bahan bacaan seperti surat kabar, resep, pedoman TV, kalendar, majalah dan buku teka teki.
- Berikan dorongan kepada anak Anda untuk membacakan kepada saudara dan temannya.
- Biar anak Anda melihat Anda menikmati membaca.
- Minta anak Anda membantu menulis daftar berbelanja, kartu selamat, kartu pos dan pesan telepon.
- Berikan dorongan kepada anak Anda untuk membaca dan mengikuti petunjuk untuk permainan, membuat mainan atau memasang peralatan baru di rumah.
- Bermainlah mainan kata dan dengarkan cerita, puisi dan lagu yang direkam, di rumah maupun di mobil.
- Bicaralah tentang plot dan akhir cerita; diskusikan alternatif dan 'bagaimana kalau'.
- Buatlah bait baru dari puisi dan lagu favorit, atau versi baru cerita dewa-dewi.

Buku yang baik untuk dibaca

Jika Anda sedang mencari buku lain bagi anak Anda untuk dibacakan di rumah, ada daftar buku yang sesuai pada halaman 10-11 dari buku ini. Carilah buku-buku ini di perpustakaan atau toko buku setempat.

 Buku permulaan dengan bab

 Buku dengan bab yang lebih panjang

 Buku lanjutan dengan bab

 Buku bergambar untuk pembaca sekolah dasar tengah

 Buku non-fiksi

中文版家长信息

课外阅读材料

这本日志是您孩子全年课外阅读材料的记录。在学年内的大多数晚上，他或她都会带一本书回家阅读。课外阅读是为了鼓励家长专门安排一段时间和子女一起进行阅读。当孩子们向家长朗读时，他们能练习在学校里所学的阅读技能。经常在家阅读的孩子能够更快更轻松地学习阅读。

如何填写日志

- 翻到第14页。每周都有这样的页面需填写。
- 在顶部（Date, Title, Pages），您的孩子写下每次阅读的日期、书名和阅读的页数。
- 每周一次，您的孩子需回顾这一周的阅读情况，并完成一个关于所阅读内容的句子。（Complete this sentence ...）
- 在第三部分，您的孩子将记录一些在阅读中遇到的新词、趣词或不寻常的词语。（Write new words ...）
- 在左下部分，您应每周为您的孩子写一次评语。评语可以是：
  - I am proud of you for reading such a big book.
    读了这么厚的一本书，我真为你感到骄傲。
  - Good on you for reading on when you get stuck!
    遇到困难时能继续读下去，真是太好了！
  - Well done! 棒极了！
  - We learnt a lot about dolphins and whales this week.
    这个星期，我们学习了许多关于海豚和鲸鱼的知识。
  - You are a great recipe reader, and a great cookie baker!
    你不但菜谱读得好，饼干也做得很棒！
- 在右下部分，老师将每周为您和您的孩子写评语。
- 在24–5、36–7 、48–9 和 60–1 页上，您的孩子用涂色或填写数字来代表全年内为达到50、100、150和200个晚上的阅读里程碑而完成阅读的夜晚。

（注：各校不一定都以同样的方式来使用这些日志，有的学校可能会不用到日志中的某些部分。）

如何开始

- 每晚留出15到20分钟的时间。
- 找一个舒适安静、适于阅读的地方。
- 和孩子并肩而坐，这样你们就都能方便地看着书本。
- 倾听孩子的阅读。到现阶段，他或她可能会为您阅读篇幅较长的书本的章节了。
- 讨论你们阅读过的内容。

如何帮助您的孩子阅读

- 您需要耐心和理解。学习阅读是一件很复杂的事，需要花时间。
- 迅速翻阅一遍书本，讨论书的类型以及图片、图表或照片。

- 鼓励孩子进行尝试。
- 给孩子足够的时间来自己解决问题。
- 对孩子说：“你觉得这个词是什么？这里应该发什么音？”
- 对孩子说：“你知道其它看上去相似的词吗？”
- 对孩子说：“读下去，我们可以回头再看这里。”
- 对孩子说：“这说得通吗？看上去对吗？”
- 随着时间的推移，帮助孩子立即识别本书12和13页上所列的常用词语。
- 随着孩子变得更为独立，您可以选择听他们朗读较短的时间，然后让他们自己默读。

在家帮助孩子阅读和书写的其它方法

- 每天对孩子诵读。读不同种类的书。
- 在阅读时，讨论所读的书。
- 前往本地图书馆，鼓励孩子不但借阅自己熟悉的书籍，还尝试新的书籍和不同作者的作品。
- 阅读并讨论诸如报纸、菜谱、电视预告、日历、杂志和智力测试书等阅读材料。
- 鼓励孩子向弟妹及朋友朗读。
- 让孩子看见您阅读并从中获得乐趣。
- 让孩子帮您写购物单、贺卡、明信片和电话留言。
- 鼓励孩子阅读和遵照说明玩游戏、搭玩具或组装新的家庭用品。
- 在家中或汽车里玩文字游戏，听录音故事、诗词和歌曲。
- 讨论故事的情节和结局以及故事发展的其它可能性。
- 为喜欢的诗词和歌曲或童话故事创作新的版本。

适于阅读的好书

如果您正在为孩子寻找其它可以在家阅读的书籍，本书10–11页上有一份合适书籍的书单。您可以到本地图书馆或书店里找这些书。

 起步书籍

 篇幅较长的书籍

 难度较高的书籍

 供小学中等年级读者使用的图片书籍

 非小说类书籍

Thông tin tiếng Việt dành cho Phụ huynh

Sách mượn để đọc ở nhà

Sổ nhật ký này ghi chép những sách con em quý vị mượn về đọc ở nhà trong năm. Em sẽ đem sách về nhà đọc hầu hết mỗi tối trong niên học. Sách mượn để đọc ở nhà là cách khuyến khích phụ huynh và con em dành riêng thời gian để cùng nhau đọc sách. Trẻ em thực tập những kỹ năng đọc đang được dạy ở trường khi các em đọc cho phụ huynh nghe. Trẻ em nào thường xuyên đọc sách ở nhà sẽ học đọc nhanh hơn và dễ hơn.

Cách điền sổ nhật ký

- Lật đến trang 14. Mỗi tuần đều có những trang giống như vầy.
- Ở phần đầu (Date, Title, Pages), con em quý vị viết ngày, tựa sách và những trang đã đọc trong mỗi lần đọc sách.
- Hàng tuần, con em quý vị nghĩ lại về những gì các em đã đọc trong tuần, rồi viết một câu mô tả một điều gì mình đã đọc. (Complete this sentence …)
- Ở phần thứ ba, con em quý vị ghi một số các từ ngữ mới, hay hoặc lạ mà em đã gặp trong lúc đọc sách. (Write new words …)
- Ở phần cuối bên trái, quý vị viết ý kiến của mình về con em mỗi tuần một lần. Ví dụ của những ý kiến là:
  - I am proud of you for reading such a big book.
    Mẹ/Cha hãnh hiện là con đọc cuốn sách dày như vậy.
  - Good on you for reading on when you get stuck!
    Giỏi lắm vì con đã đọc tiếp khi bị trở ngại!
  - Well done! Giỏi!
  - We learnt a lot about dolphins and whales this week.
    Trong tuần này, chúng ta đã học được rất nhiều điều về cá heo và cá voi.
  - You are a great recipe reader, and a great cookie baker!
    Con đọc công thức nấu ăn tài lắm và làm bánh bích-qui giỏi nữa!
- Ở phần cuối bên phải, giáo viên ghi ý kiến cho quý vị và con em đọc mỗi tuần.
- Ở trang 24–5, 36–7, 48–9 và 60–1, con em quý vị tô màu hoặc điền một con số cho mỗi tối đọc sách đã hoàn tất để đạt đến điểm mốc 50, 100, 150 và 200 tối đọc sách trong một năm.

(Lưu ý: các trường có thể không luôn luôn sử dụng các sổ nhật ký giống như nhau và một số trường bỏ trống một số phần nào đó trong sổ nhật ký.)

Cách bắt đầu

- Dành ra 15 đến 20 phút mỗi tối.
- Tìm một nơi thoải mái, yên tĩnh để đọc sách.
- Ngồi cạnh con em để cả hai có thể nhìn sách dễ dàng.
- Lắng nghe con em đọc. Ở giai đoạn này , các em có thể đọc một đoạn trong cuốn sách dày hơn cho quý vị nghe.
- Thảo luận với con em về những gì quý vị đã đọc.

Cách giúp đỡ con em đọc sách

- Kiên nhẫn và thông cảm. Tập đọc sách là công việc rất phức tạp. Việc này phải học từ từ.
- Lật sơ qua cuốn sách và thảo luận về loại sách và hình ảnh, biểu đồ hoặc hình chụp.

Thông tin tiếng Việt dành cho Phụ huynh

- Khuyến khích con em 'mạnh dạn đọc thử'.
- Để cho con em có thời gian tự suy nghĩ.
- Hãy nói 'Con nghĩ xem từ này có thể có nghĩa gì? Ý nghĩa nào có thể là đúng trong bối cảnh này?'
- Hãy nói 'Con có biết từ nào khác nhìn giống vậy không?'
- Hãy nói 'Con cứ đọc tiếp và chúng ta có thể quay trở lại sau.'
- Hãy nói 'Con thấy có đúng không? Con nhìn thấy có đúng không?'
- Dần dà, quý vị hãy giúp con em nhanh chóng nhận ra những mặt chữ thường gặp được liệt kê ở trang 12 và 13 trong sổ này.
- Khi con em quý vị độc lập dần, quý vị có thể nghe em đọc một lúc, rồi để cho em đọc thầm.

Những điều khác mà quý vị có thể thực hiện ở nhà để giúp con em đọc và viết

- Đọc sách cho con em nghe mỗi ngày. Đọc các loại sách khác nhau.
- Thảo luận về cuốn sách trong lúc quý vị đọc cho con em nghe.
- Đến thư viện địa phương và khuyến khích con em mượn cả những cuốn sách quen thuộc lẫn sách mới và các loại văn của các nhà văn khác nhau.
- Đọc và thảo luận về thứ để đọc chẳng hạn như báo chí, công thức nấu ăn, tập chương trình TV, lịch, tạp chí và sách 'puzzle'.
- Khuyến khích con em đọc cho em và bạn bè nghe.
- Để cho con em thấy quý vị thích đọc sách.
- Để cho em giúp quý vị viết danh sách mua sắm, thiệp mừng, bưu thiếp và tin nhắn trên điện thoại.
- Khuyến khích con em đọc và làm theo chỉ dẫn để chơi 'game', ráp đồ chơi hoặc lắp ráp các vật dụng mới trong nhà.
- Chơi các trò chơi sử dụng từ ngữ và nghe các câu truyện, bài thơ và bài nhạc thâu âm ở nhà hoặc trong xe.
- Thảo luận về bố cục và kết cuộc của các câu truyện; thảo luận về những khả năng khác hoặc 'giả sử rằng'.
- Đặt vần thơ, lời nhạc mới cho những bài thơ và bài nhạc được ưa chuộng hoặc tự biên câu truyện cổ tích mới.

Những cuốn sách hay nên đọc

Nếu tìm thêm sách khác cho con em đọc ở nhà, ở trang 10–11 có danh sách các cuốn sách thích hợp. Hãy tìm những cuốn sách này ở thư viện địa phương hoặc hiệu sách.

 Những cuốn sách chương đầu tiên

 Những cuốn sách chương dài hơn

 Những cuốn sách chương nâng cấp

 Sách hình dành cho các em lớp giữa của bậc tiểu học

 Sách viết về người thật việc thật

Premier's Reading Challenge

Title	Author	Date

Title	Author	Date

Do you think you can read more than 40 books?

Then photocopy page 70 before you start, so you can record all your reading.

Congratulations!
In .... (year)
.... (name)
completed
....
sessions of
take-home reading.
Well done!
....
(Teacher's signature)